# Lions

## Kings of the Grasslands

By Dr. Hugh Roome

Children's Press®

An Imprint of Scholastic Inc.

Content Consultant
Shannon Borders
Assistant Curator, Heart of Africa
Columbus Zoo and Aquarium

Library of Congress Cataloging-in-Publication Data
Names: Roome, Hugh, author.
Title: Lions/by Dr. Hugh Roome.
Description: New York, NY: Children's Press, an imprint of Scholastic Inc., [2020] | Series: Nature's children | Includes index.
Identifiers: LCCN 2018029921| ISBN 9780531229910 (library binding) | ISBN 9780531239131 (paperback)
Subjects: LCSH: Lion—Juvenile literature.
Classification: LCC QL737.C23 R665 2019 | DDC 599.757—dc23

Design by Anna Tunick Tabachnik

Creative Direction: Judith E. Christ for Scholastic Inc.

Produced by Spooky Cheetah Press

Printed in Heshan, China  62

SCHOLASTIC, CHILDREN'S PRESS, NATURE'S CHILDREN™, and associated logos
are trademarks and/or registered trademarks of Scholastic Inc.

1 2 3 4 5 6 7 8 9 10 R 29 28 27 26 25 24 23 22 21 20

Scholastic Inc., 557 Broadway, New York, NY 10012.

Photographs ©: cover: Chelsea Tischler/Gallery Stock; 1: Dorling Kindersley/Getty Images; 4 map: Jim McMahon/Mapman®;
4 leaf and throughout: stockgraphicdesigns.com; 5 child silo: Nowik Sylwia/Shutterstock; 5 lion silo: Petrovic Igor/Shutterstock;
5 bottom: Anup Shah/NPL/Minden Pictures; 6 top and throughout: Lucky Creative/Shutterstock; 7: Mercury Press/Caters News
Agency; 8 tail: Benedetta Barbanti/EyeEm/Getty Images; 8-9: mareandmare/Shutterstock; 10-11: Popova Valeriya/Shutterstock;
12-13: Achim Mittler/Frankfurt am Main/Getty Images; 15: Anup Shah/NPL/Minden Pictures; 16-17: Michel & Christine Denis-
Huot/Biosphoto; 18-19: Londolozi Images/Mint Images/age fotostock; 20-21: Mercury Press/Caters News Agency;
22-23: Denis-Huot/Nature Picture Library; 25: Jami Tarris/Getty Images; 26-27: Régis Cavignaux/Biosphoto; 28-29: Joe
Vogan/Alamy Images; 30-31: Paul Souders/Getty Images; 33: Dave King/Getty Images; 34-35: Ingo Schulz/Getty Images;
37: Look and Learn/Bridgeman Images; 38-39: Xinhua/eyevine/Redux; 40-41: Big Cat Rescue/Barcroft Images/Getty Images;
42 top: Sylvain Cordier/Biosphoto; 42 center: Eric Isselee/Shutterstock; 42 bottom: Svetlana Foote/Shutterstock; 43 top left:
courtneyk/iStockphoto; 43 top right: Michael Duva/Getty Images; 43 bottom left: Tim Ridley/Getty Images; 43 bottom right:
Freder/iStockphoto; 46: Michael Duva/Getty Images.

◀ **Cover image
shows a young
male lion yawning.**

# Table of Contents

**Fact File**.................................................................4

CHAPTER 1    **The King of Beasts**.................................6
             Built to Hunt...........................................8
             Two Types of Lions.................................11
             At Home on the Savanna........................12

CHAPTER 2    **Life in the Pride**..............................14
             Cat Chat..............................................17
             Face-Off...............................................18
             On the Hunt.........................................21
             The Lion's Share..................................22

CHAPTER 3    **A Lion's Life**...................................24
             Welcome, Cubs!....................................27
             Family Reunion....................................28
             Life Lessons........................................31

CHAPTER 4    **Ancient Ancestors**...........................32
             Big-Cat Cousins...................................35

CHAPTER 5    **Lions and Humans**...........................36
             Under Pressure.....................................39
             Looking Out for Lions...........................40

**Lion Family Tree**...............................................42
**Words to Know**.................................................44
**Find Out More**.................................................46
**Index**.............................................................47
**About the Author**.............................................48

# Fact File: Lions

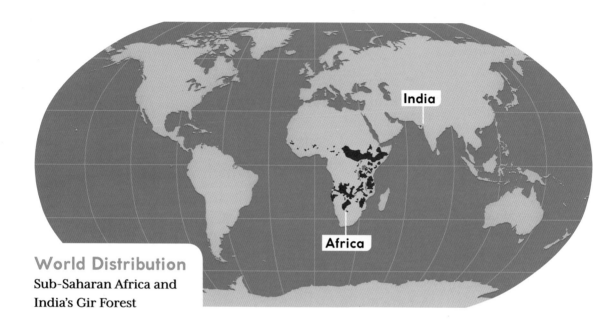

India

Africa

## World Distribution
Sub-Saharan Africa and
India's Gir Forest

## Habitat
Found mainly in
the grasslands and
forests of Africa
and India

## Habits
Live in large family
groups called
prides; females
do the majority
of the hunting

## Diet
Zebras, antelope,
buffalo,
wildebeests,
giraffes

## Distinctive Features
Very large,
muscular cats
with tan-colored
coats, long tails,
and large jaws
and paws; adult
males have manes
(dark fur around
their necks)

## Fast Fact
A lion's roar can
be heard 5 mi.
(8 km) away.

## Average Size

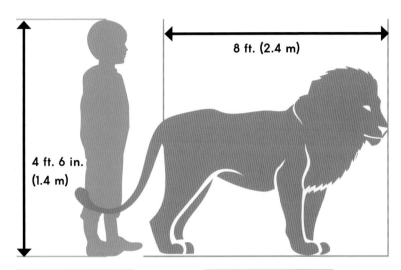

4 ft. 6 in. (1.4 m)

8 ft. (2.4 m)

Human (age 10)

Lion (adult male)

## Classification

**CLASS**
Mammalia
(mammals)

**ORDER**
Carnivora
(carnivores)

**FAMILY**
Felidae
(cats)

**GENUS**
*Panthera*
(lions, tigers, leopards
and other big cats)

**SPECIES**
*Panthera leo*
(lions)

**SUBSPECIES**
*Panthera leo
melanochaita*
(African)
*Panthera leo persica*
(Asiatic)

◀ A group of lions
has found someting
interesting in this tree.

# CHAPTER 1

# The King of Beasts

A family of lions, called a pride, lies almost hidden in the tall golden grasses. The adult females doze while several babies, called cubs, scamper around them. The cubs pretend they are hunting. They are getting ready to take their place as the **apex predator** on the **savanna**. The cubs pounce on bugs and chase each other. One reaches out to bite the tail of a large male that is snoozing nearby. Before long the little cub is joined by his siblings. They climb over the massive lion, pulling on his mane and swiping at his nose. Suddenly the male sits up and lets out a bone-rattling roar. The sound travels for miles across the still plains. This powerful hunter, who at the moment is dripping with cubs, is letting everyone know that he is at the top of the **food chain**. He is the king of beasts.

**Fast Fact**
Prides of up to 26 lions have been seen in South Africa.

▶ **This huge lion puts up with a mock attack by a playful cub.**

# Built to Hunt

Lions are powerful **carnivores** that hunt large animals for food. They are the second-largest cats on the planet. Only tigers are bigger. A healthy male lion can measure about 4 feet (1.2 meters) tall at the shoulder and over 8 ft. (2.4 m) long not including its tail. And he can weigh more than 500 pounds (226.8 kilograms) of solid muscle. Females are built similarly to males, just on a smaller scale. Lionesses grow up to 5 ft. (1.5 m) long and can weigh 400 lb. (181.4 kg). The major difference between males and females is that males have a mane, a big tuft of hair surrounding their heads.

**Fast Fact**
Lions are the only cats with tufts at the ends of their tails.

**Tail**
can grow to over 3 ft. (0.9 m) long.

**Ears**
turn in all directions
for excellent hearing.

**Eyes**
can see six
times as well as
humans' eyes.

**Jaws**
can open up
almost to 1 ft.
(30.5 cm).

**Mane**
shows this is
a male.

**Claws**
are razor sharp
and grow to 3 in.
(7.6 cm) long.

**Paws**
are the size of
dinner plates.

# Two Types of Lions

There are about 20,000 lions living in the wild today. Most of them live in Africa. However, there is another subspecies that lives in India. Asiatic lions can be found in Gir National Park.

Asiatic lions are very similar to their African cousins. They have just a few physical differences. Asiatic lions are slightly smaller than African, and they have a fold of skin that runs lengthwise along their stomachs. Male Asiatic lions have darker manes than African lions, but the fur is not as full.

Asiatic lions were once close to **extinction**. Now about 500 individuals live in protected areas in India. Though they live on a different continent, Asiatic lions' **habitats** and habits are similar to those of African lions.

◀ Asiatic lions have thicker tufts of hair on their elbows than African lions do.

11

**Fast Fact**
Tanzania has
more lions than
anywhere else
in the world.

## At Home on the Savanna

The majority of lions on Earth are found in the hot, dry savannas of southern and eastern Africa. This habitat features mostly tall grasses, some shrubs, and scattered trees. The temperature rarely falls below 60°F (15.6°C), so there are only two seasons: wet and dry. In the dry season the temperature can hover around 100°F (37.8°C). That's why most of the animals—including lions—aren't very active during the day. In fact, these gigantic cats spend up to 20 hours a day sleeping or resting. They hunt at dusk and in the early morning hours. Still, there's no better place on Earth for these hunters to live. The savanna is also home to hundreds of thousands of **prey** animals.

▶ Tree climbing is unusual for lions, but some prides have gotten into the habit of resting in trees!

# CHAPTER 2

# Life in the Pride

Lions are the only social cats in the world. They live in large family groups called prides. Prides vary in size; African prides average around 13 members. They are usually made up of females who are all related to each other: mothers, aunts, sisters, and daughters. Most African prides have only one or two adult males, known as residents. These adult males have joined an established pride. They are not related to the adult females.

Every lion in the pride has a job to do. The females, called lionesses, are the hunters. They also take care of the cubs. Males are responsible for protecting the pride and their **territory**.

In some parts of the savanna, a single pride may control an area over 100 square miles (259 square kilometers). But if the pride can find a place with plenty of water and prey, their territory might be as small as 8 sq. mi. (20.7 sq. km).

▶ The pride, young and old, move together.

## Cat Chat

It's probably not surprising to learn that such social animals have a lot of ways to communicate. Lions often greet each other by licking each other and rubbing their foreheads, faces, and chins together. They might say hello by humming or making huffing sounds. Lionesses will call to their cubs by grunting softly. The cubs respond in kind to let their moms know where they are!

Of course, lions have ways of showing displeasure, too. A pride member who's misbehaving may get a swipe from another's paw or even a nip on the neck. When a lion feels threatened, it will stand tall, hunch its back, and lift its tail to seem bigger. It will also show off its fiercest weapons: its teeth. When a lion lets out a full-throated roar, it's letting everyone within earshot know where it is, and sending a warning: *Stay off my turf!*

◄ A lion may grunt softly to check in with other members of the pride.

# Face-Off

Lions also communicate nonverbally. Holding a territory that contains reliable water and plenty of prey is key to a pride's survival. Pride males are constantly advertising how big their territory is—and how it is being protected. Roaring is one way to do this. Another is by marking the territory. Males have a **gland** between their toes that leaves a scent on the ground as they walk. They also spray urine on trees, shrubs, and grass throughout their territory. These smells warn other lions to keep away.

It is important that males defend their territory—and their pride—for obvious reasons. But females also have a strong motivation for keeping other lions away. If a new lion takes over the pride, he will immediately kill all the cubs. Still, males come and go. . . . It is the lionesses that hold the pride's territory throughout the years.

▶ Male lions may fight to protect their territory—and hold on to their pride.

# On the Hunt

Because lionesses are smaller than males, they are faster and more agile. That also makes them better hunters. Lionesses prey upon some of the biggest animals on the savanna, including buffalo, wildebeests, zebras, and antelope. So how does a 400-lb. (181.4-kg) lioness take on a 1,500-lb. (680.4-kg) buffalo? Teamwork. First the lionesses sneak up on a **herd** of prey animals. Crouching low and moving slowly and quietly, they make their way closer to the herd. They look for animals that are young, sick, or weak or that have wandered away from the rest. Suddenly a lioness springs forward and begins chasing the animal. It looks like it's getting away. Did the lioness ruin the hunt? No! She's chasing the animal directly toward her pride mates. One after another, the lionesses jump on top of the animal, biting it and digging into its **hide** with their sharp claws. Then one of them delivers the killing blow—taking the animal's throat or **snout** between its teeth. The pride will eat well tonight!

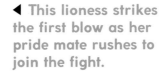

◀ This lioness strikes the first blow as her pride mate rushes to join the fight.

# The Lion's Share

The hunt didn't last very long, but it was exhausting. Lions can run at a top speed of 35 miles per hour (56.3 km per hour). But they can't keep it up for long. And wrestling a huge buffalo to the ground takes a lot of strength.

The lionesses lie on the ground, panting, as the pride males and cubs approach. The males didn't work for this meal, but they still get to eat first. After the adults have eaten their fill, the cubs will get their turn at the carcass. Lionesses are very good hunters, but their success rate isn't great. Only one out of every four attempted hunts ends in a feast. Killing large prey helps keep the pride going in between meals! But lions don't eat only what *they* kill—they are more than happy to steal meat from other animals, too.

▶ The adults have eaten their fill; now it's the cubs' turn at the zebra carcass.

# A Lion's Life

**Female lions are ready to start a** family by the time they are four years old. A lioness may go into **heat** at any time of year. She will usually **mate** with one of her pride males. Males with the fullest, darkest manes attract the most females.

To attract the male's attention, the lioness moves her tail in the air. She rubs against the lion and lies on the ground near his feet. As the lioness walks, she sprays her scent behind. The male lion sniffs the air several times. Then he opens his mouth wide, pulling his lips away from his teeth. This allows the female's scent to drift over a special odor-detecting **organ** on the roof of the lion's mouth. That is how he knows the female is ready to mate. The lion and lioness spend about a week together. In about 110 days, the lioness will be ready to give birth.

▶ A lion and a lioness nuzzle together on the savanna.

**Fast Fact**
Cubs can't roar
until they are
two years old.

# Welcome, Cubs!

The lioness moves away from the pride when she is ready to give birth. She finds a hidden spot where she can make a **den**. Lionesses usually have one to four cubs in a **litter**.

The newborn cubs are tiny. They are only 7 to 9 inches (17.8 to 23 centimeters) long and weigh 3 to 4 lb. (1.4 to 1.8 kg). The cubs are born blind and don't open their eyes until they are about one week old. They begin walking by the time they are 10 days old. The cubs are truly helpless, yet their mother must leave them alone while she hunts. Lions are **mammals**, which means they make milk for their babies to drink. If the lioness doesn't eat, she can't feed her cubs. During this time, it's crucial that the cubs stay hidden. If other predators, like hyenas or leopards, find the cubs, they will kill them. The lioness moves her den several times a month. She carries each cub, one at a time, by holding its head and neck in her mouth.

◀ The lioness uses
her massive jaws to
delicately carry this
tiny cub.

# Family Reunion

When the cubs are from six to eight weeks old, their mother rejoins the pride. Lionesses in a pride often have cubs around the same time. The cubs will all grow up together. They will even **nurse** from different lionesses if their mother isn't around. The cubs drink milk for the first two months of their lives. Then they start to add meat to their diet. By the time a cub is seven months old, its mother will no longer have milk to feed it.

There is safety for the cubs within the pride. But life on the savanna always comes with danger—even for the top predator. A lost or abandoned cub can be killed by another predator or even trampled by a buffalo. And if new males take over the pride, they will kill the very young cubs. This will ensure that only their children will become new members of the pride. After giving birth, a lioness won't be ready to mate again for 18 months. That is, unless her cub is killed. Then she can go into heat immediately.

▶ Several cubs fight for a spot to drink from this lioness.

**Fast Fact**
Some males in
Tsavo, Kenya, don't
have manes.

# Life Lessons

Young cubs spend their days playing. They pretend to **stalk** each other, insects, even tufts of grass! They are practicing their hunting skills. When the cubs are about one year old, they are ready to go along on a hunt. At first they hang back to watch and learn. By the time they are two years old, the females are able hunters and the males will have the beginnings of their manes. The male lions will now be forced out of the pride—either by the resident males or even the females. Most males join together in a **coalition**, which gives them a better chance of survival. These males will spend one to three years as nomads— living without a pride. Then they will be old enough, and strong enough, to try to take over a pride of their own.

◀ When cubs play, they are really practicing important survival skills.

# CHAPTER 4

# Ancient Ancestors

Olduvai Gorge in Tanzania, East Africa, is a famous **archaeological** site. Many amazing discoveries have been made there. In 1960, **paleoanthropologists** Mary and Louis Leakey discovered evidence of humans' first ancestors in Olduvai Gorge. That is also where the oldest recognizable lion **fossils** were found. They are 1.4 million to 1.2 million years old.

Scientists say there is little difference in the bodies of these cats and their modern-day relatives. According to research, these ancient lions spread throughout the world. Lion fossils from 600,000 years ago have been found in Italy. Even older fossils have been found in England. And some of the oldest pieces of art in existence are French cave paintings of lions from 32,000 B.C.E.

It's amazing to think that lions could once have been found on almost every continent. Today they struggle to survive in their shrunken ranges.

▶ A lion's skull hasn't changed much over the years.

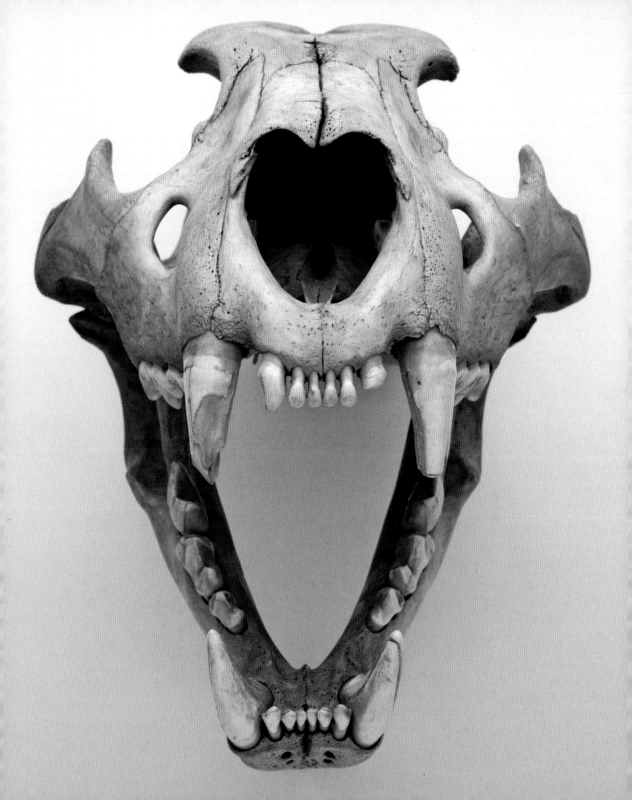

# Big-Cat Cousins

Lions belong to a group of animals known as big cats, which includes lions, tigers, jaguars, and leopards. All big cats are hunters, and all of them except lions are **solitary**. These other cats live alone except when mating or when a female has cubs.

Experts say lions closest relatives are jaguars and leopards. The jaguar is the largest cat found in the Americas. These cats are excellent climbers—and they love to swim! Like jaguars, leopards are very at home in the trees. In fact, they often spot prey from the branches. These powerful hunters also drag their kill high into a tree to protect the meat from **scavengers**. In Africa, leopards share much of the same habitats as lions. They can also be found in parts of Asia. Leopards are a threat to lion cubs, but they are no match for a full-grown lion.

◀ This leopard has spotted something and looks ready to pounce!

35

CHAPTER 5

# Lions and Humans

For more than 2,000 years, lions have been an important part of human history. They are a symbol of strength and courage. Ancient kings across the globe made the lion a symbol of their leadership. They also hunted these magnificent beasts.

In ancient Egypt, lions were considered sacred and had a city, Leontopolis, that was dedicated to them. Priests in that city fed the lions that lived in the temple—and even sang to them as they ate!

Even today, the lion is the national symbol of several countries. A constellation of stars and a sign of the zodiac are even named Leo, another name for the lion.

How, then, did these magnificent animals end up in such a vulnerable position? In the last three generations, the number of wild lions has dropped by almost half. And people are to blame.

▶ Ancient kings in the Middle East showed their bravery by hunting lions with bows and arrows.

**Fast Fact**
"Simba" is the
Swahili word
for "lion."

# Under Pressure

Over the past 50 years, the number of people in Africa
has tripled. As a result, lions' access to land and food has
decreased. When prides can't roam to hunt, they look
for other food sources. To a lion, a villager's cow is just as
good as a zebra. To protect their livestock, farmers shoot
or even poison lions. And some people hunt the same
animals lions prey on. Those lions who escape the farmers'
guns and poison can still face starvation.

Sadly, some people also shoot lions for sport. They'd
rather have a dead lion as a trophy than see these majestic
beasts roaming the grasslands.

◀ A ranger collects
the remains of lions
killed by villagers.

# Looking Out for Lions

Luckily, people and organizations from all over the world are working to preserve the population of lions living in the wild. In Africa, some governments have made it illegal to shoot lions. And farmers can face punishment if they are caught harming lions.

Rangers in Africa's game preserves protect lions from **poachers**. Some people kill the big cats for their teeth and claws, which may then be made into jewelry. In India, the government keeps hunters and land developers away from the few remaining Asiatic lions in the Gir Forest.

The battle is not over, though. It's up to all of us to make sure the lions' mighty roar is never silenced.

▶ Veterinarians in Tampa, Florida, operate to save a lion.

# Lion Family Tree

Lions belong to the felidae family, also known as cats. All the animals in this family are carnivores that hunt other animals. They all have a common ancestor that lived about 14 million years ago. This diagram shows how lions are related to other cats, such as leopards, jaguars, and tigers. The closer together two animals are on the tree, the more similar they are.

**Lynxes**
medium-sized cats with furry paws for walking on snow

**Mountain Lions**
large American cats that usually hunt alone

**Cheetahs**
slender cats that run faster than any other land animal

**Ancestor of all Cats**

*Note: Animal photos are not to scale.*

**Leopards**
spotted big cats
that often stalk
prey from trees

**Lions**
big cats that
usually live in
groups called
prides

**Tigers**
the largest cat
species; known
for their dark
stripes and reddish
fur coats

**Jaguars**
the largest of
South America's
big cats

# Words to Know

A ........... **apex** *(AY-peks)* at the top of the food chain, with no natural predators

**archaeological** *(ahr-kee-uh-LAH-ji-kuhl)* having to do with the distant past

C ........... **carcass** *(KAR-kuhs)* the dead body of an animal

**carnivores** *(KAHR-nuh-vorz)* animals that eat meat

**coalition** *(koh-uh-LISH-uhn)* a group formed for a common purpose

D ........... **den** *(DEN)* the home of a wild animal

E ........... **extinction** *(ik-STINGK-shun)* the condition of having died out

F ........... **food chain** *(FOOD chayn)* an ordered arrangement of animals and plants in which each member feeds on the one below it

**fossils** *(FAH-suhls)* bones, shells, or other traces of animals or plants from millions of years ago, preserved as rock

G ........... **gland** *(GLAND)* an organ in the body that produces or releases natural chemicals

H ........... **habitats** *(HAB-i-tats)* the places where an animal or plant is usually found

**heat** *(HEET)* the time when a female animal is ready to mate

**herd** *(HURD)* a large group of animals

**hide** *(HIDE)* the skin of an animal

L ........... **litter** *(LIT-ur)* a group of animals born at the same time to one mother

**M** .......... **mammals** *(MAM-uhlz)* warm-blooded animals that have hair or fur and usually give birth to live babies; female mammals produce milk to feed their young

**mate** *(MAYT)* to come together to produce young

**N** .......... **nurse** *(NURS)* to drink milk from a breast

**O** .......... **organ** *(OR-guyn)* a part of the body that has a certain purpose

**P** .......... **paleoanthropologists** *(pay-lee-oh-an-thru-PAH-luh-jists)* experts who use fossils and other remains to study the origin and ancestors of humans and apes

**poachers** *(POHCH-uhrz)* people who hunt or fish illegally on someone else's property

**predator** *(PRED-uh-tuhr)* an animal that lives by hunting other animals for food

**prey** *(PRAY)* an animal that is hunted by another animal for food

**S** .......... **savanna** *(suh-VAN-uh)* a flat, grassy plain with few or no trees

**scavengers** *(SKAV-uhn-jerz)* animals that eat dead and decaying material

**snout** *(SNOUT)* the long front part of an animal's head, which includes the nose, mouth, and jaws

**solitary** *(SAH-li-ter-ee)* not requiring or without the companionship of others

**stalk** *(STAWK)* to hunt or track a person or an animal in a quiet, secret way

**T** .......... **territory** *(TER-i-tor-ee)* an area that an animal or group of animals uses and defends

# Find Out More

**BOOKS**

- Blewett, Ashlee Brown. *Lion Rescue: All About Lions and How to Save Them* (Mission). Washington, D.C.: National Geographic Kids, 2014.

- Gagne, Tammy. *Lions: Built for the Hunt*. Mankto, MN: Capstone Press, 2016.

- Hatkoff, Craig. *Cecil's Pride: The True Story of a Lion King*. New York: Scholastic, 2016.

- Joubert, Beverly, and Dereck Joubert. *Face to Face with Lions*. Washington, D.C.: National Geographic, 2010.

- Walker, Sarah. *Eyewonder Big Cats: Open Your Eyes to a World of Discovery*. New York: Dorling Kindersley, 2014.

To find more books and resources about animals, visit:
## scholastic.com

# Index

**A**

African lions .................... 9, 11, 40

ancestors ........................... 32

Asiatic lions .................... *10*, 11, 40

**C**

carnivores............................ 8

claws ................................ 9

coalitions............................31

communication ........... 4, 6, *16*, 17, 18

conservation efforts............... 40, *41*

cubs....6, *7*, 16, 18, 22, *26*, 27, 29, *30*, 31

**D**

dens .................................27

diet............................22, *23*, 28

distinctive features ............... *8, 9*, 11

distribution ..............11, 12, *13*, 32, 35

**E**

extinction ........................... 11

eyesight ............................. *9*

**F**

female lions .............. 8, 14, *16*, 18, 22, 24, *25, 29*

fighting..............................*19*

food chain ............................ 6

fossils ............................... 32

**H**

habitats .................... 6, 11, 12, 35

hearing............................... *9*

heat..............................24, 28

hunting ....................12, 21, 22, 31

**J**

jaguars.............................. 35

jaws ............................... *9, 26*

**L**

Leakey, Mary and Louis ............... 32

leopards .......................... *34*, 35

litters ...............................27

**M**

male lions...8, 14, 18, *19*, 22, 24, *25*, 28

mammals...............................27

manes...........6, 8, *9, 10*, 11, 24, 31, 35

mating...................... 24, 25, 28

**N**

nursing............................28, *29*

**O**

Olduvai Gorge ........................ 32

# Index *(continued)*

**P**

paws ................................................. 9

poachers ................................ 39, 40

predators ..................................... 28

prey .............. 12, 18, *20*, 21, 22, 35, 39

prides ........................ 6, 14, *15*, 18, 28

**R**

relatives ....................................... 35

roaring ................................... 6, 17

**S**

scent ...................................... 18, 24

size and weight ....................... 8, 27

speed ............................................ 22

stalking ....................................... 31

symbolism .................................. 36

**T**

tails ........................................... 8, *8*

territory ................................ 14, 18

threats .............. 27, 28, 36, *37*, *38*, 39

tigers ........................................ 8, 35

tongues ....................................... 22

tree climbing .......................... 12, *13*

**W**

whisker spots .............................. 11

# About the Author

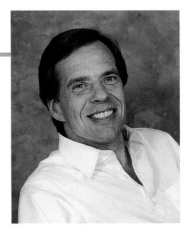

Dr. Hugh Roome is Executive Vice President of Scholastic Inc., where he was publisher of *Scholastic News*, *Science World*, and *The New York Times Upfront*. This book is dedicated to the staff of the Tufts Wildlife Clinic at the Cummings School of Veterinary Medicine, where Hugh is a member of the Board of Advisors.